WARSHIP PICTORIAL 35

UNITED STATES NAV_

TICONDEROGA CLASS CRUISERS

Author: Kurt Greiner • Editor: Steve Wiper

USS SHILOH (CG-67)

Author's Dedication: *To my brothers, Karl, Greg, Jim and especially Frank, who was like a second father to me.*

CLASSIC WARSHIPS PUBLISHING
P. O. Box 57591 • Tucson, AZ. 85732 • USA
Web Site: www.classicwarships.com • Ph/Fx (520)748-2992

ISBN 978-0-9823583-6-8
Printed by Arizona Lithographers, Tucson, Arizona

The Ticonderoga Class Guided Missile Cruisers

The Ticonderoga class guided missile cruisers represent the most capable class of anti-aircraft warships ever built. Designed to beat the threat of Soviet saturation missile attacks, their advanced war-fighting systems have proven flexible enough to face the threats of the new millennium. Though the path from design to commissioning was not an easy one, their superior performance has proven an invaluable asset to the Navy today, and into the future.

Aegis – The Shield of the Fleet.

The Kamikaze attacks of the Second World War had a profound impact on the United States Navy. The difficulty of destroying an enemy willing to press home an attack regardless of personal danger spurred the development of advanced anti-aircraft gun systems, and later, ship-borne guided missiles. The Navy made a heavy investment in this new technology. The building and converting of entire ships without guns in the late 1950's showed a serious commitment. While the missiles of the time were somewhat effective at destroying enemy aircraft and missiles, it was soon realized that ships could be overwhelmed by a determined assault, especially if coordinated from multiple directions and using higher speeds provided by advances in aeronautics.

The first system developed to counter the threat of mass attack was an advanced, analog based system called Typhoon, but despite great efforts, it was not effective against all the missiles of it's time, and was never deployed. The idea behind it, an ability to assess and prioritized targets using radar and computing became the core of the Advanced Surface Missile System (ASMS), which evolved into Aegis.

Aegis, in mythology the shield of Zeus, was designed as a complete war-fighting system, integrating sensors and weapons to provide quick reaction time and accurate assessment and prioritization of threats. Again drawing on an earlier, less successful technology, it incorporated phased array radar, the SPY 1, and a computer running what was, at the time, one of the most complicated software codes ever written. If this system could be brought to the fleet it would represent a tremendous increase in protection against missiles and aircraft

Finding the right ship for the system.

Getting Aegis deployed was to prove a tremendous challenge due to fiscal and political considerations during its development in the late 1960's and early 1970's. The United States Navy was facing the block obsolescence of many ships built during and just after World War II. Tremendous numbers of warships were due to be retired, and the high cost of new construction prevented replacement on a one for one basis.

Admiral Elmo Zumwalt, the Chief of Naval Operations at this time, sought to implement a system called the "*high-low*" mix, which would create of fleet of highly capable destroyers (which became the DD-963 SPRUANCE Class), and less expensive escorts (eventually the OLIVER HAZARD PERRY FFG-7 Class Guided Missile Frigates). Although in the long run both classes cost far more than initially predicted, these ships formed the backbone of the destroyer force for two decades.

The policy was not without it's critics. The PERRYs were derided as inferior ships, with too many compromises in their design, and the large Spruance destroyers were considered under armed for their size. The FFG's proved to be very useful, and have had longer service lives than many of their contemporaries. A more careful examination of SPRUANCEs showed many advantages not immediately obvious – excellent hanger facilities, large weapon magazines, superior guns, and most importantly, a margin for growth built into the design that was to prove most useful later.

During this same time period there was a group of naval officers, lead by Admiral Hyman Rickover, who were pushing very hard for expensive, nuclear powered escorts. Rickover, who had established control of the US Navy's nuclear program, saw additional nuclear powered surface ships as a means to expand his power base and provide projects for his design teams. With his powerful political connections Rickover intended to limit the introduction of Aegis, if at all, to a few large ships, promoting designs for atomic powered strike cruisers and other budget breakers. At best, only a handful of these ships could be built, which meant that there would not be enough Aegis equipped ships to protect all the aircraft carriers and surface action groups.

Much infighting followed, and at one point, it seemed that Aegis might be cancelled, as it was a expensive system itself that seemed unlikely to find a complimentary ship design. Fortunately, far sighted officers in the Aegis planning office had quietly been working on plans to integrate the system into an existing, proven hull, that of the Spruance class, which ultimately proved to be the salivation of the project. The choice of the hull was a fortunate one, as the SPRUANCE had reserves of space and buoyancy that would be sorely needed for the new design. The final go ahead was given and construction began. Since the system was to be built using a destroyer hull, she was given the next hull number in the guided missile destroyer sequence, DDG-47 (DDG-46 was the USS PREBLE, last of the COONTZ class).

As Michael C. Potter points out in his excellent book, *Electronic Greyhounds*, the Aegis system was built around modular deckhouses that were extremely heavy. In fact the system added topweight equivalent to three triple six inch turrets, such as those used in the CLEVELAND class light cruisers, but with the disadvantage of not having the barbettes, armor or magazines of those ships to help lower the center of gravity. Much thought was given to reducing weight wherever possible, but in the end, some safety margins had to be cut, again only possible because much data had already been collected from SPRUANCES in the fleet. Even then, when the first ship of the class was inclined to test her stability, she was found to be severely overweight, and more compromises accepted. Again, the robust design of the DD-963 made this possible, as it was found the hull was stronger than expected and therefore could be overloaded without excessive danger. The extra weight makes the ships ride lower than the SPRUANCES and a raised bulwark was added to the Fo'c'sle to keep the ships drier. Despite heavy reinforcement these have been damaged or even lost in heavy weather.

By the time the TICONDEROGA was launched by First Lady Nancy Reagan, it was designated a cruiser, and the designator changed from DDG, to CG. Her number was left unchanged, which explains the gap in the Navy's cruiser numbering sequence (MISSISSIPPI, CGN—41 was the last cruiser built prior to the TICONDEROGA).

Turbine powered and built for speed

While a cruiser, the TICONDEROGA class has propulsion system and rudders of her destroyer cousins. She is powered by 4 General Electric's LM-2500 Gas Turbines of roughly 21,000 HP each, which are very similar to the engines used on DC-10 aircraft. They can run on DFM (Diesel Fuel Marine), although they also have the ability to use JP-5, the standard aviation fuel used by the US Navy. These turbines run through gears to twin shafts, reducing the speed down to 55 to 168 RPM dependent upon throttle settings. Variable pitch propellers allow both direction control (as turbines cannot be reversed) as well as fine control over the lower speed range. These propellers also provide a useful capability to slow the ship quickly in what is termed a "*crash stop,*" by reversing the blade pitch while they spin, which causes rapid deceleration. The propeller blades have passages in them for compressed air (the "*Prairie*" system) to form a curtain of noise reducing bubbles to reduce and mask the sound signature of them (the hull is equipped with the similar "*Masker*" system to reduce hull machinery noise).

The TICONDEROGAs are equipped with dual rudders, and can turn very sharply, so much so that those with expertise in handling these ships suggest handling limits. A combination of speed and rudder angle not exceeding 30 is recommended to avoid excessive listing while turning. Witnessing one of these ships engage in a high speed maneuver is an exciting, if somewhat anxious sight, as they tend to lean out of the turn and return upright a bit slower than other classes when the helm is centered.

Armament

The primary weapons of the CG-47 class are her missiles. The first five units carried two of the twin armed Mk. 26 missile launchers, and therefore could fire four missiles at one time. Although there were plans to modernize these ships by replacing those units with VLS (Vertical Launch System), it was decided to retire them instead, leaving only the VLS ships remaining. Each VLS has space for 64 cells in a 8 by 8 matrix, although three cells are lost to a folding crane incorporated in the unit. Plans are underway to delete this crane, which cannot handle the heavier Tomahawk missiles, thereby increasing the number of cells to 128

The capacity and flexibility of the VLS system makes the TICONDEROGA's one of the most powerfully armed ships of all time. Their ability to launch dozens of Tomahawk BGM-109 missiles (currently, the theoretical number is 122, though in practice Standard Missiles are carried for air defense) allows them to strike at targets further inland than any gun system. The Tomahawk can deliver a selection of warheads, including, at one time, nuclear. Since the US Navy does not comment on the presence of nuclear weapons on it's ships, it is unknown if any of the ships of this class carried that warhead before policy was changed and that weapon taken out of active service. The conventional payload of the missile packs an incredible punch – a thousand pound warhead or a load of bomblets. They can be guided to the target by DSMAC - Digital Scene Matching Area Correlation, where an image of the projected route is stored in the missile and constantly compared to the terrain as the weapon passes over. More recent versions are equipped with GPS as well, though the Military is sensitive to the vulnerabilities of that system. It has a range of 1500 miles and a speed of 550 miles per hour.

In the Fleet Air Defense role, TICONDEROGA's carry the Standard Missile. The current version is the SM-2 Block III, a highly capable weapon against threats from the air. Depending on the variant, it has a range of from 40 to 90 miles at Mach 3.5. The Standard is a smart missile that does not need continuous guidance to find a target, giving the TICONDEROGA's the ability to engage 20 targets simultaneously using it's four directors. In addition to the basic anti-air role, the SM-2 has a surface mode that can be used to attack ships. Some units of the class can carry the Standard SM-3, an extended range (270+ mile) weapon used for Theater Ballistic Missile Defense and in one case a satellite.

For shorter-range strikes against surface targets, the cruisers mount the RGM-85 Harpoon missile in self contained canisters located on the main deck near the fantail. Up to a maximum of eight can be carried in two racks of four, but the number has varied in recent years as threats have changed. The Harpoon has a range of over 75 miles when launched from a ship, flies over 500 miles per hour and can carry a warhead weighing approximately 500 pounds.

There are future plans to include the RIM-162 Evolved Sea Sparrow Missile in at least some of the VLS cells. These can be carried 4 to each VLS cell (the "*quad pack*"), and are intended to provide quick, reliable point defense against incoming missiles.

The main gun armament of the ship is comprised of two Mk. 45 5"/54cal. automatic gun mounts, one on the bow and the other located on the main deck at the stern. This mount has a 20 round magazine that can be fired without reloading by the crew. If additional rounds are needed, the magazine can be replenished while the gun is firing. It can shoot a 70-pound shell to a range of approximately 15 miles.

To fire at small surface targets, the ships are equipped with the Mk. 38 Bushmaster, a 25mm chain gun that is manually aimed. It can fire single shots, or in automatic mode 100 or 200 shots per minute. Each round is a little over one pound in weight, and can be fired over two miles. Some later versions of this mount are stabilized. At first these weapons were only temporarily installed on ships being deployed to the Middle East, but plans are in place to install two mounts per ship on a permanent basis near the aft VLS.

The last ditch gun defense against air threats is the CIWS (Close In Weapon System), also known as the Phalanx. This is a 20mm Gatling type gun in a totally integrated package that includes radar, magazine and electronics. The radar tracks both incoming targets and the outgoing rounds, attempting to merge both tracks to assure destruction of the threat. Recently, an upgrade has been introduced to allow the engagement of surface targets using Forward Looking Infrared (FLIR), current plans are to upgrade all the CIWS to this Block 1B configuration. The complexity of these mounts has resulted in reliability problems, leading wags in the fleet to say CIWS stand for "*Christ it won't shoot!*" Modifications to decrease downtime are being implemented.

Lighter machine guns are increasingly being mounted for port security, especially since the attack on the USS COLE (DDG-67). These consist of single and more recently, dual .50 cal. machine gun mounts Typically the guns are mounted on the bow, along the 01 level, and elsewhere on the ship as needed. Light arms for boarding parties and pier side security are also carried aboard.

The ships mount two Mk. 32 triple torpedo tubes for use in Anti-Submarine Warfare. These can fire lightweight Mk. 46, 50 or 54 Torpedoes. Unlike other classes of ships, the SPRUANCE and the TICONDEROGA classes carry the tubes inside the ship, behind a sliding door visible on both the port and starboard sides of the hull under the flight deck. The preferred scenario for antisubmarine warfare is to kill submarines outside their range, using standoff weapons or helicopters, but these torpedoes are carried in case a submarine is discovered close in, to allow a rapid shot. Quick response to a detected enemy is vital, as a submarine maneuvering wildly to avoid an incoming torpedo is likely to loose the ability to accurately control it's own wire controlled weapons.

Electronics

The SPY-1A (and later 1B) phased array radar is the heart of the Aegis system. The four faces, each composed of a hexagon with unequal sides, are mounted on the face and side of the forward and after superstructure, allowing 360° coverage. There are no moving parts. Each of the 4480 radiating elements found in each face is electronically controlled, and can be grouped as needed to provide wide coverage or pencil beams to track many targets at once. The radar is frequency agile and can select different modes as needed to avoid or reduce jamming. The advanced signal processing of the Aegis system allows targets to be tracking near the sea surface, allowing the engagement of sea skimming missiles designed for radar avoidance.

Beyond SPY-1A, the ship is equipped with the SPS-49 air search radar, a fully stabilized, two-dimensional bearing and range finding unit. It acts as a backup to the SPY-1A, and can also be used by itself when it is tactically advantageous to mask the presence of the Aegis ships distinctive emissions, as the SPS-49 is carried by a variety of US Navy and NATO ships. The SPS-49 rotates at 6 or 12 RPM.

For surface search and navigation the SPS-55 radar is carried, with a bar shaped antenna 6 feet across.

Four SPG-62 illuminators are carried aboard TICONDEROGA class cruisers, as opposed to three in the ARLEIGH BURKE class destroyers, and can be identified by their 7.5 foot diameter dish. These radar guide the SM-2 during the final flight phase to the target.

The SPQ-9A is both a surface and low level (less than 2000 feet) air search radar housed in a distinctive 10 foot by 8 Foot radome, and is tied into the gunfire control system.

Electronic Warfare is the role of the ubiquitous SLQ-32 ("*Slick 32*") found on so many US naval ships. The CG-47 class carries the V(3) version, which adds the capability to engage in active jamming. It can be seen as a box-like structure on high up on the ships, on both sides. It's purpose is to receive the signal of enemy radar, including missile radar, analyze the bearing, and initiate counter-measures, including the deployment of chaff from the mortar like SRBOC (Super Rapid Blooming Chaff) launchers.

The SRBOC is being supplemented on many surface warships by an Australian designed decoy called Nulka. This is a rocket-propelled unit that can hover using rotors and move away from the ship at a constant height and speed while repeating back missile radar, presenting an attractive target for incoming missiles. Although the exact flight time is classified, sources report it is at least 55 seconds, far longer than conventional chaff remains effective, particularly in adverse weather situations.

To detect submerged targets, the CG-47 class ships are equipped with the bow mounted SQS-53, capable of both active and passive modes of operation. Starting with CG-54, the towed array, SQR-19 TACTAS (Tactical Towed Array Sonar) is installed at the stern, and allows the ship to listen for submarines at a distance that reduces interference by it's own noise. The 3 in. cable is 5600 feet long, and provides not only acoustical data, but also information on temperature, depth and heading of the array, which can be controlled to depths over 1000 feet. The ability to change array depth gives the surface ship the opportunity to exploit the thermocline, or area of rapid change in temperature that can affect sound propagation, and reduce the advantage the submarines held in sound detection for many years.

Service Life

Five of the original TICONDEROGAs have been retired. At this writing, the TICONDEROGA (CG-47) herself is being held for use as a possible museum, and VALLEY FORGE (CG-50) was sunk in a Fleet Training Exercise in late 2006. The other three, YORKTOWN (CG-48), VINCENNES (CG-49) and THOMAS S. GATES (CG-51) have been stricken and are awaiting disposal.

VINCENNES was involved in the shoot down of Iran Air Flight 655.

PRINCETON (CG-59) was damaged by a mine explosions during the first Gulf War on 18 February 1991. She was mended in the Middle East, and then proceeded home for permanent repairs.

LAKE ERIE (CG-70) successfully shot down a satellite in decaying orbit on 21 February 2008.

Photographed entering San Francisco Bay for Fleet Week 1996, the USS VINCENNES (CG-49) cruises behind a fireboat shooting water out of it's monitors in celebration. The VINCENNES was one of the first 5 TICONDEROGA class cruisers, all of which have been de-commissioned.
Photo by Lee Upshaw / SeaPhoto

Above: The USS PHILIPPINE SEA (CG-58) in June, 1995. Even from a distance these ships are quite recognizable due to their distinctive superstructures.
Photo courtesy of Leo Van Ginderen via SeaPhoto.

Right: One of the last units of the class, the USS PORT ROYAL (CG-70) departs the Sydney, Australia harbor in April 1997. The PORT ROYAL made history in July 1993 as the first U.S. Navy ship to be commissioned in Hawaii.
SeaPhoto collection.

The bow of the USS PORT ROYAL (CG-73) February 2009 in dry dock, at Pearl Harbor, HI. She is being repaired after a grounding earlier that month. Note the dark rubber window on the sonar dome, which allows efficient transmission of sound. *US Navy Photo*

The guided-missile cruiser USS LAKE CHAMPLAIN (CG 57) drops anchor February 2009. The bulwark seam is clearly visible near the upper edge of the hull. Note that the TICONDEROGA class only carries a bow and starboard side anchor, and that they are Stockless anchors. *US Navy Photo*

A bow view of the USS SAN JACINTO (CG-56), taken July, 1993 at Wilhelmshaven, Germany. Based on the SPRUANCE class destroyer hull, the TICONDEROGA class cruisers have had a bulwark added at the bow to reduce the wetness forward.
Photo by Frank Findler

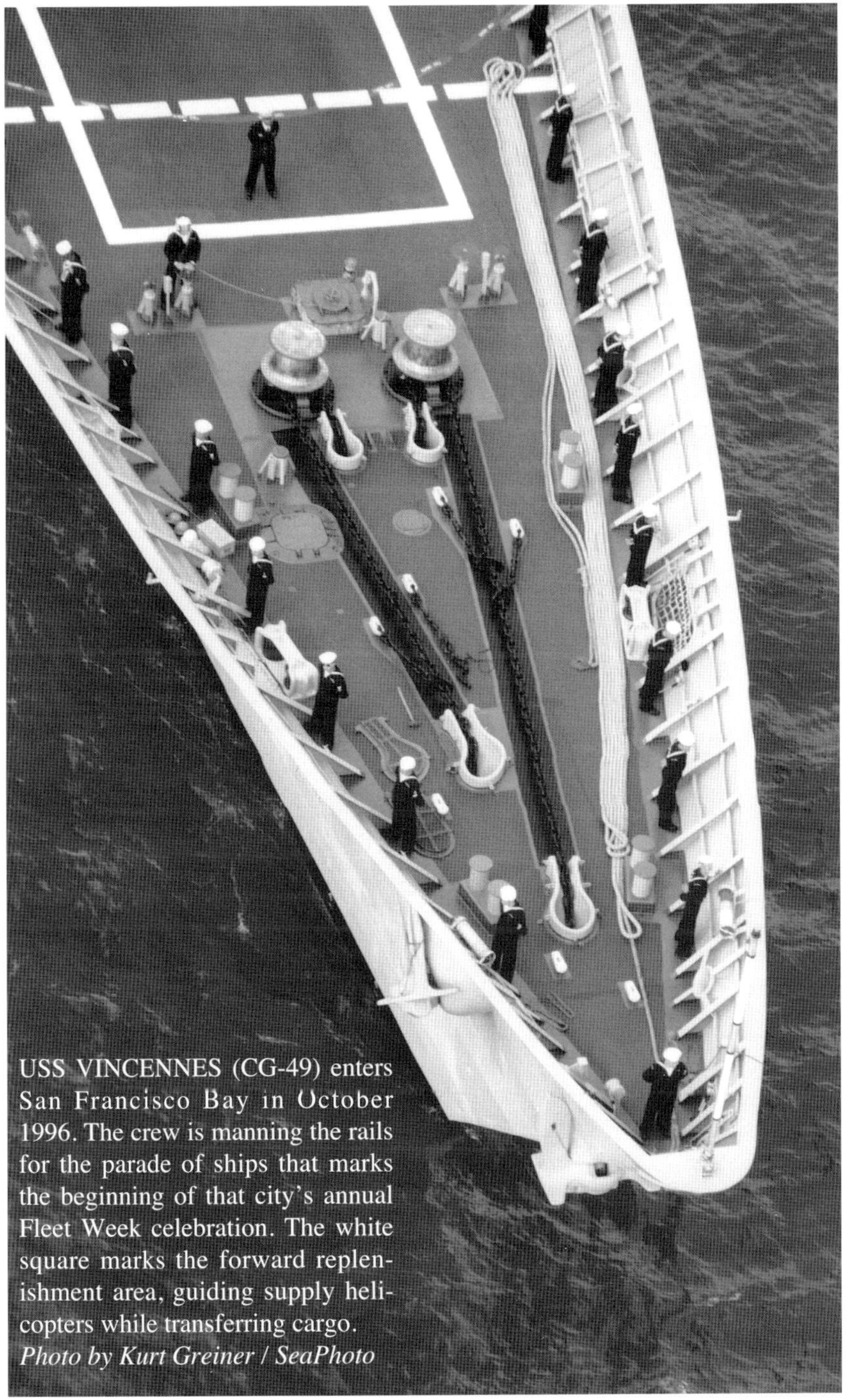

USS VINCENNES (CG-49) enters San Francisco Bay in October 1996. The crew is manning the rails for the parade of ships that marks the beginning of that city's annual Fleet Week celebration. The white square marks the forward replenishment area, guiding supply helicopters while transferring cargo.
Photo by Kurt Greiner / SeaPhoto

A more oblique view of the USS SHILOH (CG--67). While in general USN ships had their hull numbers painted in light gray and ocean gray after 1997, the SHILOH is clearly an exception in this October of 2003 photo.
Kurt Greiner / SeaPhoto

The USS MOBILE BAY (CG-53) departs San Francisco Bay in October 2003. The white square marks the forward replenishment area, guiding supply helicopters while transferring cargo.
Photo by Kurt Greiner / SeaPhoto

A closer view of the ground tackle aboard the USS ANTIETAM (CG-54), as well as the mooring gear. The shields forward of the hydraulic controls protect the operators from flying debris when anchoring. Hard hats and gloves are also mandatory during this operation. The top of the hand wheels on the controls for the anchor windless and the hawsepipe covers were painted green for starboard and red for port.
Kurt Greiner / SeaPhoto

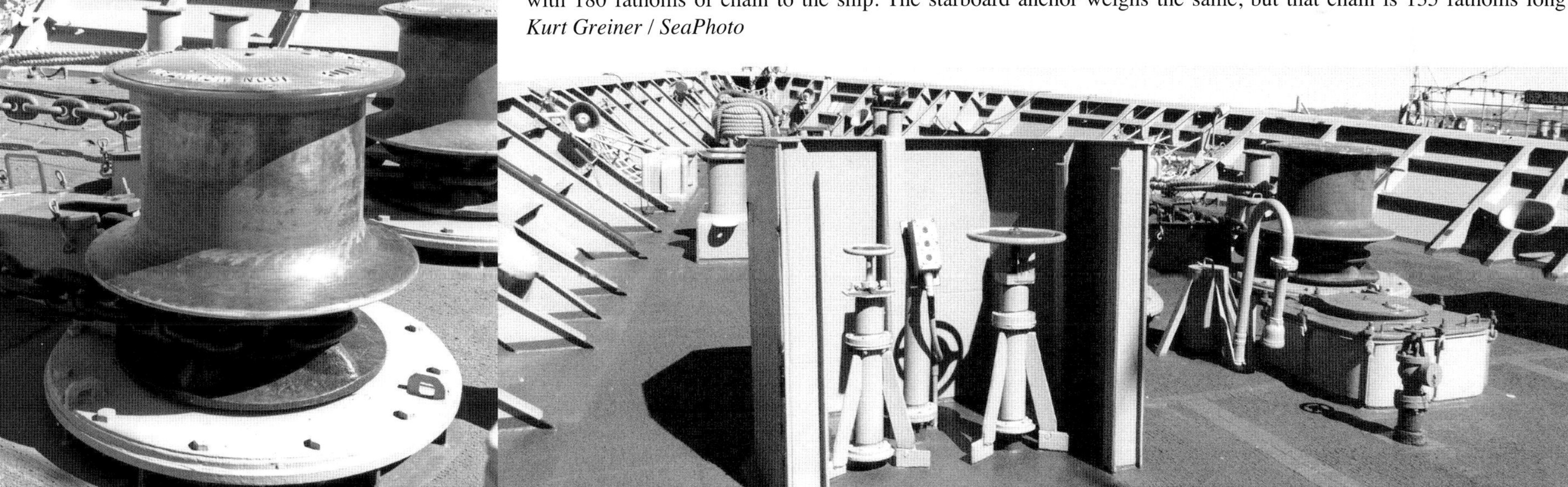

The port capstan control on board the ANTIETAM. The capstan lifts the 1632 pound bow anchor, which is attached with 180 fathoms of chain to the ship. The starboard anchor weighs the same, but that chain is 135 fathoms long. *Kurt Greiner / SeaPhoto*

Left: The forward 4" gun and Mk. 26 Missile launcher on the USS VINCENNES (CG-49), October 1996. The twin arm launcher was installed in the first five units of the class.

Above: The forward Mk. 41 missile launcher (VLS) on board the USS PRINCETON (CG-59), October 2006. This system is simpler and more reliable than the earlier Mk. 26 twin arm launchers.

Photos by Kurt Greiner / SeaPhoto

The USS CHANCELLORSVILLE (CG-62) entering San Francisco Bay in the early 1990's. Note the lack of satellite communication domes when compared to later photographs.
Photo by Kurt Greiner / SeaPhoto

The USS VALLEY FORGE (CG-50) in 1999. Note on the bridge wings the various letters "E" and "H". The White "E" with a shadows indicates that the ship has won a battle efficiency competition; others are for excellence in various operations. See page 40 in the color section for more information.
Photo by Kurt Greiner / SeaPhoto

Left: The bridge area of the USS LAKE CHAMPLAIN (CG-57), August 2002. The hexagon shaped object is the SPY-1 radar. The dish to the right with holes in the back is an OE-82 satellite communication antenna.
Kurt Greiner / SeaPhoto

Below & Right: The bridge of the USS PHILIPPINE SEA. Note 1 - The video camera, aimed at the forward VLS cells to monitor missile launches safely, and 2 - The circular devices on the bridge windows are glass discs that spin, using centrifugal force to repel water.
SeaPhoto Collection

Below: The port forward superstructure of the USS PHILIPPINE SEA (CG-58). Note the piping for the decontamination system, which allows the ship to rinse itself in the event of nuclear, or chemical attack. *SeaPhoto Collection*.

Above: Looking aft of the bridge area on the USS PORT ROYAL (CG-73) in 1999. Circled is the starboard side Phalanx close-in weapon system (CIWS), which provides last-ditch defense against missiles and other threats. *Kurt Greiner / SeaPhoto*

Above: A view of the starboard side, just aft of the foremast, on the USS LAKE CHAMPLAIN (CG-57). The ready service ammunition lockers for the CIWS mounts are located on the deck between them. *Kurt Greiner / SeaPhoto*

Midships view of the USS PHILIPPINE SEA (CG-58) in 1995. Note that the motor whaleboat and associated handling gear have been replaced with Rigid Hull Inflatable Boats (RHIB's). *SeaPhoto Collection*

Right: USS SHILOH's port side forward superstructure, June 1997. The circled area is one of the ships refueling stations. Note the encapsulated life rafts on racks just above and aft. *Kurt Greiner / SeaPhoto*

The USS MOBILE BAY (CG-53) in October 2000. Compared to earlier views, note the RHIB storage and plethora of satellite communication domes.
Kurt Greiner / SeaPhoto

Left: Amidships area USS SHILOH (CG-67) in 1997. Note the forward and aft stacks are offset port and starboard. This was a characteristic of the SPRUANCE class destroyers from which the TICONDEROGA's were derived.
Kurt Greiner / SeaPhoto

Above: A view of the mainmast of the guided missile cruiser USS TICONDEROGA (CG-47) showing the quadruple legged mast of the first two units, TICONDEROGA and YORKTOWN (CG-48), later units had tripod masts.
US Navy Photo

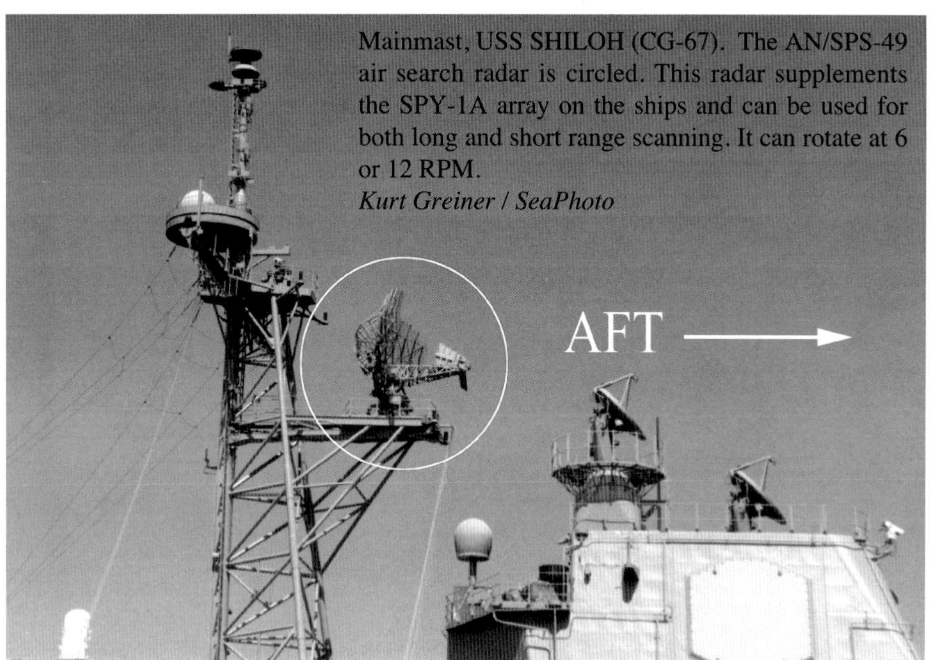

Mainmast, USS SHILOH (CG-67). The AN/SPS-49 air search radar is circled. This radar supplements the SPY-1A array on the ships and can be used for both long and short range scanning. It can rotate at 6 or 12 RPM.
Kurt Greiner / SeaPhoto

Main mast, USS ANTIETAM (CG-54), 1997. The upper circle identifies the TACAN (Tactical Air Navigation) system, with a lighting rod above it. In the lower circle is the SPS-64 navigation radar.
Kurt Greiner / SeaPhoto

Below: The AN/SPS-49 Radar aboard USS SHILOH. This radar is carried by a wide variety of ships, thus allowing the Aegis cruisers the option of searching in a mode that doesn't readily identify them via their distinctive SPY-1A array, which can be tactically useful.
Kurt Greiner / SeaPhoto

Right: Mainmast foundation on board the USS CHANCELLORSVILLE (CG-62). Note the construction details, utilizing flat plates where the support tubes join with the main structure. *Kurt Greiner / SeaPhoto*

Right: Mainmast head, USS SHILOH. The arrow points to the circular ring of the OE-120/UPX, part of the UPX-29 IFF (Identification Friend or Foe) system. *Kurt Greiner / SeaPhoto*

After superstructure, USS VINCENNES, 1996. The stern would be to the right in this photo. The dishes are the AN/SPG-62 Fire Control Radar, which illuminates the target for missile guidance. *Kurt Greiner / SeaPhoto*

The aft superstructure of the USS VINCENNES (CG-49), 1996. Sailors man two saluting guns (circled) just forward of the mainmast. The aft AN/SPG-62 Fire Control Radar (arrow) aboard the VINCENNES. A red circle surrounds the lower radar to indicate the danger zone when the dish traverses. *Kurt Greiner / SeaPhoto*

Left: An inboard view of the port stacks on the USS VINCENNES (CG-49). Note how the soot discolors the long range antennas on the outboard side.
Kurt Greiner / SeaPhoto

Below Left: A side view of the AN/SPG-62 Fire Control Radar on board the USS VALLEY FORGE (CG-50). Open source publications quote the peak power of this narrow band radar at about 10 kilowatts.
Kurt Greiner / SeaPhoto

Below Right: Starboard side aft superstructure of USS VALLEY FORGE (CG-50) 2002. The exhaust stacks for the gas turbines are prominent in the foreground.
Kurt Greiner / SeaPhoto

Helo Deck, USS VINCENNES. The safety nets are rigged down for flight operations. Note that the SH60B's are stored at a slight angle (nose pointing to port) on the flight deck.
Kurt Greiner / SeaPhoto

Upper Right: In this photo, the two tracks for the RAST (Recovery Assist, Secure and Traverse) system are visible under the SH60B helicopter. This device allows recovery in rougher conditions by winching the helicopter down onto a moveable platform that is then pulled inside the hanger. There is one track for the port, and another for the starboard hanger.
Kurt Greiner / SeaPhoto

Left: An overhead photograph showing the helicopter deck markings on the USS PRINCETON (CG-59) in 2006. The dotted "T" line is for larger helicopters when they perform Vertical Replenishment; if the rotor hub is kept on or behind that line they will clear obstructions on the ship. *Kurt Greiner / SeaPhoto*

Below: The flight deck of the VALLEY FORGE. The circled area on the hull shows where the Mk. 32 triple torpedo tubes are located behind pneumatically operated doors. *Kurt Greiner / SeaPhoto*

Left: Thick glass protects the helicopter control station on the starboard side of the flight deck. This provides a safe place to direct flight operations from the ship.
Kurt Greiner / SeaPhoto

Below: A view from the starboard side of the flight deck, USS LAKE CHAMPLAIN (CG-57). A refueling station, one of four on board, is in the circled area.
Kurt Greiner / SeaPhoto

An aft quartering view of the USS CHANCELLORSVILLE (CG-62), showing two of the SPY-1A faces. The flight deck safety nets are rigged in their normal configuration.
Kurt Greiner / SeaPhoto

USS COWPENS (CG-63) in the early 1990's. Note she carries a conventional Captain's gig on the port side, and a RHIB on the starboard.
Kurt Greiner / SeaPhoto

Above: The area above the hanger includes helicopter landing aids, such as the Stable Glide Slope Indicator (circled). *Kurt Greiner / SeaPhoto*

Hanger Door Detail, USS VELLA GULF (CG-72). This 2009 photo shows still more satellite domes on the superstructure, and the access hatches in the hanger doors when they are closed.
Photo by Lee Upshaw/ SeaPhoto

Below: A look inside the open hanger of the USS VINCENNES (CG-49). It can hold 2 SH60 Helicopters. The first two units, TICONDEROGA (CG-47) and YORKTOWN (CG-48) originally carried the smaller SH-2F, but were later upgraded.
Kurt Greiner / SeaPhoto

Looking up at flight deck level from the aft missile deck on the USS VELLA GULF (CG-72). Note that the ladder location can vary from ship to ship. A white painted ready storage locker for the chain gun is on the left side. The thin panels help reduce the heat inside. *Lee Upshaw / SeaPhoto*

The aft area of the USS CHANCELLORSVILLE (CG-62) in the early 1990's. The deck has a relatively fresh coat of anti-skid coating and looks darker than normal. This coating fades to a much lighter shade after a few months.
Kurt Greiner / SeaPhoto

The 01 Level (Missile Deck) and the 02 Level (Flight Deck) on the USS LAKE CHAMPLAIN (CG-57) in 2002. This Mk. 41 Vertical Launching System (VLS) can launch 61 missiles in a very short time.
Kurt Greiner / SeaPhoto

USS VINCENNES & IRAN AIR FLIGHT 655

The USS VINCENNES (CG-49) is perhaps one of the most well known units of the TICONDEROGA class guided missile cruisers due to it's involvement in what became known as *"The Vincennes Incident"* on July 3, 1988.

This occurred during the Operation PREYING MANTIS, when units of the US Navy were deployed in the Persian Gulf to insure the flow of oil despite the war then taking place between Iran and Iraq. Ships of several nations were re-flagged as US merchant vessels, and escorted by US warships to ensure their safety.

In the year prior to the shoot down, two USN ships had been severely damaged in the course of these duties. The SAMUEL B. ROBERTS (FFG 57) detonated an Iranian mine in April of 1988, and the STARK (FFG 31) by 2 Exocet missiles fired by Iraq in May of 1987. The STARK had been caught using *"lax procedures"* and several of her officers were punished as a result. Both ships were nearly lost, but good damaged control prevented either ship from sinking.

It was into this environment that the VINCENNES, commanded by William C. Rogers, sailed in June of 1988, following a request by the Task Force Commander for an Aegis equipped warship to guard against air attacks. On June 18, an alert was issued warning naval units to be on guard against aggressive behavior.

On the morning of July 3, VINCENNES was patrolling near Abu Musa, roughly halfway between Abu Dhabi in the United Arab Emirates and Bandar Abbas in Iran. She received word that Iranian Boghammer speedboats had attacked a freighter, and sent her helicopter to investigate, following at high speed. The speedboats fired on the helicopter, and after some back and forth with command, VINCENNES was allowed to pursue them into Iranian waters. She exchanged fire with them for several minutes, until a misfire silence her forward 5" gun. So that she could bring her aft gun to bear, Rogers ordered sharp maneuvers to keep that mount unmasked. This resulted in severe heeling, throwing objects around the CIC and causing some confusion.

At roughly the same time,. IRAN Air flight 665, an Airbus A-300 airliner carrying 290 passengers and crew, took off from Bandar Abbas, an airfield used by both military and civilian aircraft. An error by a sailor in the CIC resulted in misidentification of the flight as a military aircraft, and further errors showed the aircraft as descending, not climbing, leading Captain Rogers to think this was a F-14 attacking his ship. Rogers ordered VINCENNES to fire two SM-2 missiles from her forward launcher, and both impacted the airliner at 13,500 feet. There were no survivors. The US has since expressed regret over the incident and provided compensation to the victims families.

The entire incident took 7 minutes, from take off to missile launch, and is an example of the pressure a modern military commander is under to make rapid decisions involving life or death. Had the aircraft been an F-14, VINCENNES may well have been hit by anti-ship missiles, just like the STARK. Iran's rhetoric before and after the incident made a *Kamikaze* attack a viable possibility. While there is no doubt that tragic mistakes lead to the firing of the missiles, it would be unfair to dismiss the environment of tension generated by these *"just short of war"* operations, or Iran's role in maintaining this tension.

The USS VINCENNES (CG-49) in 2003, two years before her de-commissioning. *US Navy Photo*

This is a view showing a modified Standard Missile 2 (SM-2) Block IV interceptor launched June 5, 2008 from the guided-missile cruiser USS LAKE ERIE (CG 70) during a Missile Defense Agency test to intercept a short-range ballistic missile target. The missiles intercepted the target approximately 12 miles above the Pacific Ocean 100 miles west of Kauai, Hawaii on the Pacific Missile Range Facility.
US Navy Photos

Left: An SH-60B off the USS VELLA GULF takes off for flight operations. The sled for the RAST (Recovery Assist, Secure and Traverse) system is circled on deck. This system allows safer recovery by winching down the helicopter from a hover.
US Navy Photo

Above: A SH-60B Seahawk Helicopter on board the USS CHOSIN (CG-65). The Seahawk is designated LAMPS (Light Airborne Multipurpose System) III. While a potent Anti-submarine weapon, it can also be used for Search and Rescue, Gunfire Support, Anti-ship attack, and Vertical Replenishment,
Kurt Greiner / SeaPhoto

Above: A SH-60B on board the USS VALLEY FORGE (CG-50). The streamers hanging down in various places are remove before flight tags, indicating protective devices that must be removed, or critical systems that must be activated before flight.
Kurt Greiner / SeaPhoto

Above Right: Note the hinged tail section on this SH60B, on board the USS VALLEY FORGE (CG-50).
Kurt Greiner / SeaPhoto

Right: SH60B on board the USS ANTIETAM (CG-54).
Kurt Greiner / SeaPhoto

Left: The stern of the COWPENS, showing her "*cattle horns*" painted next to the name. A few ships in the navy have had stern names that were done in non-traditional manner. Another example was the USS JOHN HANCOCK (DD-981) that had the signature of her famous namesake instead of regulation block lettering. *US Navy Photo*

The guided missile cruiser USS Cowpens (CG 63) near the completion of it's Ship's Repair Force (SRF) dry dock period in Yokosuka, Japan, 2004. Note the black colored rubber coated window on the sonar dome. *US Navy Photo*

Above: The USS PORT ROYAL (CG-73) preparing to leave drydock at Pearl Harbor in September, 2009. The ship had been damaged by grounding in February of that year, and required extensive repairs.
US Navy Photo

Left: A stern view of the PORT ROYAL. The blue color anti-fouling paint is a new silicone based, non-toxic coating developed by Naval Sea Systems Command. The slick surface is projected to save significant fuel and maintenance costs by impairing the ability of marine organisms to attach themselves.
US Navy Photo

This 2005 photo of the USS CHANCELLORSVILLE (CG-62) shows her plowing through heavy seas prior to an underway replenishment. The stress the bow bulwark has to take is very evident; hence it's extensive bracing. *US Navy Photo*.

Aft Superstructure, USS SHILOH in 1997, as she passed under the Golden Gate Bridge, entering San Francisco Bay, California.
Kurt Greiner / SeaPhoto

The USS VINCENNES in 1996, about to pass under the Golden Gate Bridge, San Francisco, California.
Kurt Greiner / SeaPhoto

The USS VALLEY FORGE (CG-50) enters San Francisco Bay, October 1999.
Kurt Greiner / SeaPhoto

The TICONDEROGA class guided missile cruiser USS VINCENNES (CG 49) steams away from USS KITTY HAWK (CV 63) after an underway replenishment (UNREP) evolution, early in the morning.
U.S. Navy Photo

The USS LAKE ERIE (CG 70) fires a Harpoon anti-ship missile during a maritime exercise. The Harpoon is launched from tubes on the fantail instead of the VLS. The capabilities of this missile have improved over the years, offering various types of terminal attack (sea skimming or pop up and dive) and mission profile (direct or using waypoints, to hide the launching ship location) *US Navy Photo*

Top Left: The USS NORMANDY (CG 60) fires a 5 inch 54 caliper MK 45 canon. The gun can train at 30∞ per second to ± 170∞, and elevate at 20∞ per second from -15∞ to +65∞.
US Navy Photo

Right: Tracer rounds are fired from a MK. 38 Mod. 2 25 mm machine gun system aboard the USS LAKE CHAMPLAIN (CG 57) during a live-fire exercise. This is the newest, remotely controlled version of the Bushmaster system.
US Navy Photo

Left: Sailors launch an inflatable gunnery target balloon, or "*killer tomato,*" from the guided-missile cruiser USS ANZIO (CG-68) during a live-fire exercise. This particular model features moderate radar reflectivity and costs about $ 700.
US Navy Photo

Above: The helm of the USS VELLA GULF (CG-72). Note the twin throttles for the gas turbine engines (circled) instead of the traditional engine telegraph. The pitch of the propellers is also controlled from this station.
Duane Curtis photo

Right: Crew members monitor radar screens in the combat information center aboard the guided missile cruiser USS VINCENNES (CG-49). Each station covers a different threat zone or weapon system, while the Captain can maintain overall situational awareness using the large screens.
US Navy photo

Top Left: Bridge level, USS PRINCETON (CG-59) in 2004. A signal lamp is in the foreground. Even in the modern Navy, there is a place for signal lamps and flags when the ships are under EMCON (Emissions Control) when they want to avoid detection by passive electronic means. Above: Signal Shelter, USS PRINCETON.
US Navy Photos/Classic Warships Collection

Right: Starboard side flag bag, USS PRINCETON (CG-59). The ship's signal flags are stowed inside on racks to allow quick access.

Left: View to starboard of platform adjacent to the flag box.
US Navy Photos/Classic Warships Collection

A color view of the USS MOBILE BAY (CG-53). Note small plank of varnished wood on the top of the bridge wings.
Kurt Greiner / SeaPhoto

Above Right: Combat ribbons and awards on the USS CHANCELLORSVILLE (CG-62), 2004. The gold shaded "E" is for winning the battle completions five years in a row; winning once results in a white "E" with black shading. Other color "E"'s are black for maritime warfare, red for engineering, blue for logistics, yellow for safety, and purple for efficiency, and green for command and control. The "H" is a health award for crew wellness. *US Navy Photo*

Left: Underway Replenishment (UNREP) station on board the USS VELLA GULF (CG-72). One of 4 such stations on the ship, they allow the rapid transfer of the NDF (Naval Distillate Fuel).
Lee Upshaw photo

Right: Crew members of the guided missile cruiser USS LEYTE GULF (CG 55) crane a Rigid Hull Inflatable Boat (RHIB) back aboard ship. The RHIB and associated handling gear are much lighter than the conventional boats they replaced, an advantage with these top heavy ships.
US Navy Photo

A view from on board the SACRAMENTO class Fast Combat Support Ship, USS SEATTLE (AOE 3), showing the USS HUE CITY (CG 66) during an Underway Replenishment (UNREP) exercise in support of Operation ENDURING FREEDOM. *US Navy Photo*

A view of the port side, forward superstructure of the USS MOBILE BAY (CG-53) in 1995. Note the location of the squadron or group badge. This can be a Cruiser / Destroyer Group, or a Carrier Strike Group.
US Navy Photo / Classic Warships Archive

Another view of the MOBILE BAY portside, this time the after section of the warship. Note the Captain's gig has been replaced by a RHIB in this 1995 photo.
US Navy Photo / Classic Warships Archive

TICONDEROGA Class Variations

Unlike the BURKE class guided missile destroyers, which are divided into *"Flights,"* the TICONDEROGA class are divided by *"Baseline."* Note that many ships have been upgraded since their initial construction to a higher Baseline level.

Baseline O
The first two ships of the class, CG-47 and 48, had the Mk. 26 twin missile system, first version of Aegis, SPY-1A radar, SH-2F LAMPS I helicopter. No towed sonar arrays. The mainmast had 4 legs. 05 Level forward of the mainmast.

Baseline 1
CG-49 through 51. Mk. 26 twin arm missile, first version of Aegis, SPY-1A radar, SH-60B LAMPS III helicopter. RAST helicopter haul down system, helo data link. Tripod masts on these and all later ships. 05 Level deleted. 04 level extended to back legs of mainmast.

Baseline 2
CG-52 through 58. Introduced VLS in place of Mk. 26 twin arm launchers, which gave the ships the ability to fire Tomahawk cruise missiles. CG-53 and all later substitute a Rigid Inflatable Boat (RIB) for the 26' Motor Whaleboat. CG-54 and later have towed sonar arrays. HY-80 steel for the hull introduced was first introduced on these ships as a weight saving measure, as was the lowering of the aft Aegis deckhouse by two feet, and the shortening of the 04 level to just aft of the CIWS.

Baseline 3
Hull numbers CG-59 through 67. Lighter, more capable SPY-1B replaces SPY-1A, which allows these and later units to be used for Theater Ballistic Missile Defense

Baseline 4
Hull numbers CG-68 through 73. Improved computers, incorporating 32 bit processing versus 16 bit in prior units. Upgraded sonar capability.

Special thanks to Walt Briese for assistance in preparing this section.

The USS MOBILE BAY (CG-53) steams though the North Arabian Gulf in 2008. Serving on these ships is not for the faint of heart - or stomach.
US Navy Photo

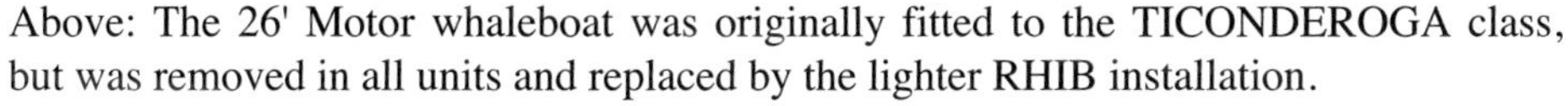

Above: The 26' Motor whaleboat was originally fitted to the TICONDEROGA class, but was removed in all units and replaced by the lighter RHIB installation.
Kurt Greiner / SeaPhoto

Below: Captain's gig, USS SHILOH. The Captain's gig always had red striped boot topping. Admiral's gigs had green striped boot topping.
Kurt Greiner / SeaPhoto

Above: The Captain's gig location on the USS SHILOH (CG-67) in 1997. These boats were carried longer than the 26' Motor Whale Boat, but were eventually replaced by the RHIB's as well.
Kurt Greiner / SeaPhoto

Below: Captain's gig aboard the USS CHOSIN (CG-65).
Kurt Greiner / SeaPhoto

Above: The first boats to be replaced on the TICONDEROGA class were the starboard side 26' motor whaleboats. Here the USS VALLEY FORGE carries her RHIB in the earlier, high mounting with large crane. *Kurt Greiner / SeaPhoto*

Below: Crew members lower a Rigid Hull Inflatable Boat (RHIB) aboard the guided missile cruiser USS VELLA GULF (CG 72). This is the later and simpler RHIB handling configuration. *US Navy Photo*

Right: Another photo illustrating the early RHIB storage scheme, this time aboard the USS LAKE CHAMPLAIN (CG-57). *Kurt Greiner / SeaPhoto*

Top Right: Encapsulated life raft storage, USS LEYTE GULF.
Lee Upshaw / SeaPhoto

Lower Right: New life raft containers were photographed on board the USS LEYTE GULF in October of 2009. They are noticeably less round than their predecessors.
Lee Upshaw / SeaPhoto

Above: The initial type of 25 man encapsulated life raft containers on board the USS Lake Champlain (CG-57). They are fitted with a hydrostatic switch to automatically deploy and inflate if the ship sinks.
Kurt Greiner / SeaPhoto

The Mining of the USS PRINCETON (CG-59)

The USS PRINCETON was the only Aegis cruiser to be damaged in combat. The incident took place early in the morning on 18 February 1991 during the first Gulf War. At the time, the cruiser was part of the naval group feinting an amphibious assault on Iraq from the Gulf.

A pressure sensitive bottom mine detonated under the cruiser, with a second mine exploding near the bow, probably from the shock of the first. As a result of the two explosions the ship was severely whipped, and suffered major structural damage, including a 6 in. crack from one side to the other of the aluminum superstructure. A good segment of that structure was separated from the deck. The combat system was knocked out due to shock damage and loss of cooling, and would take between 15 minutes and 2 hours to restore (sources disagree).

The blast damaged the port rudder, sending the ship out of control. It was finally manhandled back in place. Although the turbines could be lit off, it was decided to have the USS BEAUFORT (ATS-2) tow the ship out of the area due to uncertainty about her hull strength. BEAUFORT hauled her to Bahrain for initial, temporary repairs, and from there she spent some time in Dubai with a repair ship and a final dry docking to make her seaworthy using materials delivered by cargo plane that had been diverted from new construction. From there she sailed to Long Beach Naval Shipyard for additional repairs. Repairs to the ship took two months and cost $24,000,000. Less than a year after the attack, both sets of main reduction gears failed, probably from either the shock of the attack or a misalignment in the shafts. According to former crew members, it took several years for all the bugs to be worked out, but the ship remained an effective warship, winning several battle efficiency awards.

The USS PRINCETON (CG-59) in 2008. *US Navy Photo*

USS LEYTE GULF (CG-55) in October 2009. She was involved in a bad shipyard fire in 2007 that injured 5 industrial workers. This received wide attention at the time as it was thought it might have been a terrorist attack, but it was determined the cause was an industrial accident.
Photo by Lee Upshaw / SeaPhoto

A view of the stern of the USS MOBILE BAY (CG-53) in October 2003, as that warship was entering the San Francisco Bay, California. MOBILE BAY was the second of the VLS equipped ships. Note the mount and shield for .50 cal. machine guns, fitted port and starboard, just forward of the VLS launch tubes, at the deck edges.
Kurt Greiner / SeaPhoto

Another view of the USS MOBILE BAY (CG-53) in October 2003, this one of her helicopter pad and after superstructure.
Kurt Greiner / SeaPhoto

Continuing forward on the USS MOBILE BAY (CG-53), October 2003.
Kurt Greiner / SeaPhoto

This page is a blend of two photographs and is a testament to the quality of Kurt Greiner's photography for his business *SeaPhoto*. The two images were snapped seconds apart, but as you can see they are nearly a perfect match to each other.
Steve Wiper - Editor

USS MOBILE BAY (CG-53) Forward superstructure in this view.
Kurt Greiner / SeaPhoto

Primary Photo: The stern of the USS VALLEY FORGE (CG-50). Note the way the lines are laid out on the fantail in preparation of mooring.
Kurt Greiner / SeaPhoto

Inset Photo: The fantail of the USS VINCENNES (CG-49). The OE-416/SRC Antenna Group is circled. This is for long range communications, antennas smounted on hinges and can be folded outboard 90 degrees as needed.
Kurt Greiner / SeaPhoto

Above: USS LEYTE GULF (CG-55), Norfolk, VA 2009. Note the empty Harpoon racks on the fantail.
Lee Upshaw / SeaPhoto

Left: Aft section, USS SHILOH, 2003. In later years, the space on either side of the VLS would be used to mount Bushmaster Chain Guns.
Kurt Greiner / SeaPhoto

The USS LAKE CHAMPLAIN (CG-57) undergoing a drydock availability. With the propellers removed, the shape of the bearings and support struts can be clearly seen.
Kurt Greiner / SeaPhoto

The stern of the USS CHANCELLORSVILLE (CG-62). On the lower right are the two AN/SLQ-25 NIXIE openings. These are towed devices intended to decoy torpedoes through acoustic and electronic means. The single opening in the center is for the AN/SQR-19 Tactical Towed Array SONAR (TACTAS), a mile long cable studded with microphones to passively detect submarines. *Kurt Greiner / SeaPhoto*

TICONDEROGA CLASS SPECIFICATIONS

Length Overall	570 Feet
Length Waterline	563.6 Feet
Beam	55 Feet
Displacement, Full Load	9466 Tons (there will be variations among units)
Draft	31.5 Feet
Propulsion	4x GE LM-2500 Gas Turbine, 80,000 HP, 2x shafts, 5 bladed, variable pitch propellers, 2x rudders
Speed (Official)	30+ Knots
Range	6000 Nautical Miles at 20 knots
Crew	387 (28 Officers, 359 Enlisted), plus aviation detachment

Helicopters	Facilities for 2x LAMPS III (Earlier units upgraded to LAMPS III)
Missiles	
VLS Ships	122x Cells, various missile types.
Mk 26 Ships	2x Twin Arm launchers, 88x missiles in magazines
All Ships	Up to 8x Harpoon missiles in two deck mounted quad canisters
Guns	2x 5-inch/54 Mk. 45 2x 20mm Phalanx CIWS Bushmaster Mk. 38/Mod. 0 Recently upgrading to Mod. 2. Lighter weapons vary among units and deployment, including twin & single .50 cal. MG, + numerous LAWs.

Torpedoes	2x Mk. 32 Triple ASW torpedo tubes
Radar	4x SPY 1A/B phased array 1x SPS-49 Air Search 1x SPS-55 Surface Search 1x SPS-64 Navigation
Sonar	SQS-53 A/B/C Bow SQR-19 Towed Array
Electronics Warfare	SLQ-32 Nixie Decoy Mk 36 SRBOC Chaff Launchers Nulka Decoy (Some units)
Cost	Approximately $ 1,000,000,000 (1980's dollars)
Operating Cost	Approximately $ 25–35M per year.

A Standard Missile Three (SM-3) is launched from the guided missile cruiser USS SHILOH (CG 67) during a joint Missile Defense Agency, U.S. Navy ballistic missile flight test. Two minutes later, the SM-3 intercepted a separating ballistic missile threat target, launched from the Pacific Missile Range Facility, Barking Sands, Kauai, Hawaii.
US Navy Photo

Mk. 26 Guided Missile Launcher. The first five ships of the class carried two of these twin arm missile launchers, one forward and one aft, which had a magazine capacity of 44 missile each. They could fire the first two versions of the Standard Missile, as well as ASROC, but the inability to fire the Tomahawk Cruise Missile was a major disadvantage.
All photos Kurt Greiner / SeaPhoto

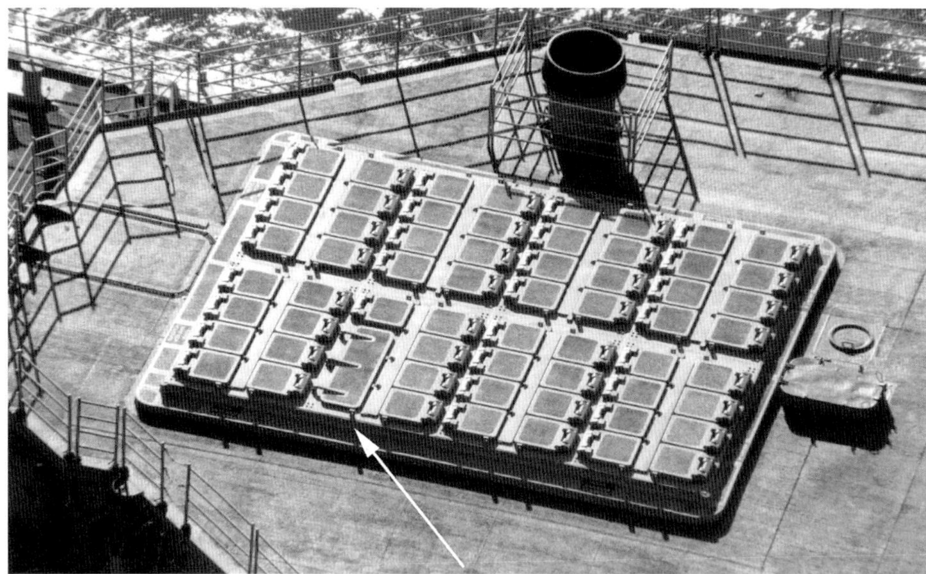

Above: The Mk. 41 Vertical Launching System (VLS). Each of the two VLS systems can carry 61 missiles (3 cells are replaced by a collapsible strikedown crane, indicated by the arrow, that can be used to reload some types of missiles). Simpler maintenance, better capacity and faster rate of fire quickly made the VLS, once available, the standard system for the fleet. *Kurt Greiner / SeaPhoto*

Above: The aft Mk. 41 VLS on board the USS ANTIETAM (CG-54). The VLS could launch the later RIM-156 Standard Extended Range Block IV, as well as the RIM-161 Standard Missile 3 (SM-3) that can intercept ballistic missiles. *Kurt Greiner / SeaPhoto*

Right: VLS on USS CHANCELLORSVILLE (CG-62). Hinged steel plates (one is indicated by the arrow) behind each block of cells flip up to vent the exhaust gas of the missile upward during a launch (see page 29). *Kurt Greiner / SeaPhoto*

Left: Inspecting the missile hatches on the USS HUE CITY (CG-66). A berthing barge, or YRBM is in the background. *US Navy Photo*

Above: Empty Harpoon Racks on board the USS VALLEY FORGE (CG-50). After the Cold War, many ships no longer carry a full complement of Harpoons at all times, particularly when not on deployment. Often they will have only two tubes per rack, or in this case, none at all. *Both Photos Kurt Greiner / SeaPhoto*

Below: A view from the port side of the CHOSIN. In 1988, Harpoon missiles were used to sink the Iranian frigate SAHAND during Operation Praying Mantis.

Above: Harpoon Launchers USS CHOSIN (CG-65). The RGM-84 Harpoon is a surface to surface missile that can travel at over 500 miles per hour to a range of over 75 miles. It can deliver a 490 pound penetrating high explosive warhead. *Kurt Greiner / SeaPhoto*

Below: Mk. 141 Harpoon Launchers on board the USS LAKE CHAMPLAIN (CG-57). Although the TICONDEROGA's carry other missiles suitable for anti-ship warfare, the Harpoon is the only system designed for that specific purpose. *Kurt Greiner / SeaPhoto*

5 in. (127mm) /54 caliber Mk. 45/Mod. 2 Automatic Naval Gun System.

This was the standard dual purpose (DP) gun mount installed on USN ships from 1973 to 1999 when it was superceded on later units by an upgraded version. It is operated by a crew of six, all below deck level, and can fire a 70 pound projectile at the rate of 16 to 20 rounds per minute with a maximum range of 25,900 yards or just under 15 miles.

This gun system can fire a variety of munitions, including point detonating fused (PDF), variable timed fused (VTF), infrared fused (IR), mechanically timed fused (MTF), electronically settable fused (ESF) and extended length rounds, such as Rocket Assisted Projectiles (RAP). *All photos this page Kurt Greiner / SeaPhoto*

Above: The forward 5"/54 caliber gun on board the USS CHANCELLORSVILLE (CG-62). Note the open access hatch which allows a rare peek inside this automated mount, and the practice rounds in front. Photographed October 1991 during San Francisco's *"Fleet Week."*
Kurt Greiner / SeaPhoto

Above & Below: The Mk. 32 Triple Torpedo Launcher is located behind doors in the hull on both the port and starboard side. They fire the Mk. 46 and Mk. 50 Anti-Submarine Torpedoes. *Kurt Greiner / SeaPhoto*

Below Left: The guided missile cruiser USS THOMAS S. GATES (CG 51) fires a training round from an Mk. 45 -5 inch/54 caliber lightweight gun during a live fire exercise in the Pacific Ocean. The crew often puts sheets of plywood or other protective coverings on the deck during firing practice to prevent ejected shells from gouging the anti-skid coating. *US Navy Photo*

Top Left: The Mk. 15 Phalanx Close In Weapon Systems (CIWS) on board the USS LAKE CHAMPLAIN (CG-57). The Phalanx was designed as the last line of defense against missiles and other targets. It can fire sabot rounds of tungsten or depleted uranium at 3000 to 4500 rounds per minute. *Kurt Greiner / SeaPhoto*

Top Right: The Guided Missile Cruiser USS PORT ROYAL (CG 73) fires a Phalanx Close In Weapons System (CIWS) during a routine maintenance exercise. *US Navy Photo*

Right: The white dome contains the search radar (top) and the tracking radar (lower). Each ammunition drum holds 990 rounds, with additional reloads nearby in ready service lockers. *Kurt Greiner / SeaPhoto*

Far Right: The CIWS is a fully automatic and integrated system. It can find, evaluated and prioritize threats, and using two radar systems, track both the incoming target and outgoing rounds, correcting to bring them together. *Kurt Greiner / SeaPhoto*

Top Right: Mk. 38/Mod. 2 Machine Gun System (MGS) on board the USS LEYTE GULF (CG-55) in 2009. The Mod. 2 mount moves control of the weapon to the Combat Information Center to reduce crew exposure. It fires 25mm ammunition at a rate of 180 rounds per minute.
Lee Upshaw / SeaPhoto

Top Left: An aft view of the Mk. 38/Mod. 2 on board the USS HUE CITY (CG-66). The system was first mounted on the USS PRINCETON (CG-59).
Duane Curtis Photo

Right: Maintenance on the Mk. 38/Mod. 2 system, aboard USS LAKE CHAMPLAIN (CG-57).
US Navy photo

Far Right: Mk. 38/Mod. 2 on board the USS HUE CITY (CG-66). *Duane Curtis photo*

Left: A M240B fired from the bridge wing of the USS CHOSIN (CG 65). This is a belt-fed, gas-operated medium machine gun firing the 7.62x51mm NATO cartridge.
US Navy photo

Below: A single .50 caliber machine gun on the USS NORMANDY (CG 60).
US Navy photo

Right: Twin 50 caliber gun mount, USS NORMANDY (CG-60).
US Navy photo

Top Right: The Mk. 36 Super Rapid Bloom Offboard Counter-measures (SRBOC) Chaff and Decoy launching system. A mortar launched decoy that is used as a last resort to lure missiles away from their intended targets by a combination of radar reflective materials and intense heat. A variety of rounds can be fired, including SRBOC, Sea Gnat and TORCH. *Kurt Greiner / SeaPhoto*

Top Left: Decoy launchers, USS VELLA GULF (CG-72). The Nulka launcher is circled; this decoy, developed in Australia, deploys rotors that allows it to hover while it radiates emissions designed to entice incoming missiles from the ship. The heat generated by it's rocket motor lures infrared guided weapons. *Duane Curtis photo*

Right: A single Nulka launcher on board the USS HUE CITY (CG 66). Nulka is from an Australian Aboriginal word meaning *"be quick!"*. *Duane Curtis photo*

Left: Another view of the decoy launchers on board the CG-72. *Duane Curtis Photo*

This photo, taken in June of 2010, shows the AN/SPQ-9B Radar antenna for that system, now carried by the USS BUNKER HILL (CG-52). On TICONDEROGA class cruisers, this radar replaces the AN/SPQ-9, it's antenna that was carried in a large radome, now giving greater capability against low threats, such as cruise missiles. The antenna rotates at 30 RPM.
Kurt Greiner / SeaPhoto

The guided-missile cruiser USS BUNKER HILL (CG 52) performs maneuvers with the littoral combat ship USS FREEDOM (LCS 1) and the Nimitz-class aircraft carrier USS CARL VINSON (CVN 70) in early 2010. *US Navy Photo*

Ship Names & Lineage

Most of the TICONDEROGA class cruisers are named after significant battles, but like virtually all postwar USN warship classes, there is an exception. The prefix USS denotes that vessel as commissioned into the United States Navy and can be used only when a vessel is in commission. USS means United States Ship.

USS TICONDEROGA CG-47
Pre-Revolutionary and Revolutionary War battles.

USS YORKTOWN CG-48
Penultimate battle of American Revolution.

USS VINCENNES CG-49
Revolutionary War.

USS VALLEY FORGE CG-50
Famous encampment during Revolutionary War.

USS THOMAS S. GATES CG-51
Secretary of Defense under President Eisenhower.

USS BUNKER HILL CG-52
Early battle of the American Revolution.

USS MOBILE BAY CG-53
Civil War Naval Battle "Damn the Torpedoes, Full Speed Ahead!!"

USS ANTIETAM CG-54
First Civil War battle in the North, with 23,000 casualties in one day.

USS LEYTE GULF CG-55
Largest Naval Battle of World War II, effectively the end of Imperial Japanese Navy.

USS SAN JACINTO CG-56
Decisive Battle of the Texas Revolution in 1836.

USS LAKE CHAMPLAIN CG-57
War of 1812, Land and Naval battle ended the British invasion of northern territory.

USS PHILIPPINE SEA CG-58
"The Marianas Turkey Shoot"; Largest carrier borne aircraft battle in history.

USS PRINCETON CG-59
Victory by George Washington in New Jersey.

USS NORMANDY CG-60
Invasion of France, June 1944.

USS MONTEREY CG-61
1846, Mexican-American war, victory by Zachary Taylor.

USS CHANCELLORSVILLE CG-62
Civil War battle, 1862, considered Lee's greatest victory.

USS COWPENS CG-63
Decisive battle of Revolutionary War, considered a tactical masterpiece.

USS GETTYSBURG CG-64
Turning point of the Civil War in the North.

USS CHOSIN CG-65
Decisive battle during the Korean War.

USS HUE CITY CG-66
American Victory during the Tet Offensive during Vietnam War, 1968.

USS SHILOH CG-67
Early Union Victory (1862) under Generals Grant and Sherman.

USS ANZIO CG-68
Invasion of Italy 1944.

USS VICKSBURG CG-69
Civil War battle gave control of the Mississippi River to the Union in 1863.

USS LAKE ERIE CG-70
Key naval victory in War of 1812; lead to the recapture of Detroit.

USS CAPE ST. GEORGE CG-71
Captain Arleigh Burke ends the Imperial Japanese Navy's "Tokyo Express" in 1943.

USS VELLA GULF CG-72
First US Navy victory in a night torpedo duel with the Imperial Japanese Navy.

USS PORT ROYAL CG-73
Union Naval Victory in South Carolina, 1861, also a Revolutionary War battle.

The USS MONTEREY (CG 61) in 2006. This is a good view of the massive superstructure of the TICONDEROGA class cruisers. *US Navy Photo*

USS ANTIETAM (CG-54). *SeaPhoto Collection*

REFERENCES

Combat Fleets of the World 2002-2003

A. D. Baker, Naval Institute Press, 2003

Electronic Greyhounds

M. Potter, Naval Institute Press, 1995

Naval Institute Guide to Ships & Aircraft of US Fleet

N. Polmar, Naval Institute Press, 17th. Ed. 2001

U. S. Destroyers - Revised Edition

N. Friedman, Naval Institute Press, 2004

World Naval Weapons Systems

N. Friedman, Naval Institute Press, 1989

RESOURCES

Federation of American Scientist

Web Site: www.fas.org

SeaPhoto

Web Site: www.warshipphotos.com

U. S. National Archives @ College Park

8601 Adelphi Rd., College Park, MD. 20740-6001
(301)713-6800 • Web Site: www.nara.gov

U. S. Naval Historical Center

805 Kidder Breese SE, Bldg # 57
Washington Navy Yard, Washington DC, 20374-5060
Still Photos (202)433-2765 • Web Site: www.history.navy.mil

U. S. Naval Institute

291 Wood Rd., Annapolis, MD 21402-5034
(800)713-8764 • Web Site www.usni.org

U. S. Navy News Stand Web Site

Web Site: www.news.navy.mil

RM ROMA will be the topic of Warship Pictorial 36.

ACKNOWLEDGMENTS

Classic Warships

would like to express it's gratitude for assistance from the following individuals -

Duane Curtis • Lee Upshaw
Frank Findler • Michael Winter
Ryan Carey • Walt Briese

WARSHIP PICTORIAL SERIES

W. P. # 4 USS Texas BB-35
W. P. #10 Indianapolis & Portland
W. P. #20 HMS Hood
W. P. #22 USS Ticonderoga
W. P. #23 Italian Heavy Cruisers
W. P. #24 Arleigh Burke Class Destroyers
W. P. #29 North Carolina Class Battleships
W. P. #30 IJN Takao Class Cruisers
W. P. #31 USS Buchanan DD-484
W. P. #32 South Dakota Class Battleships
W. P. #33 USS Lexington CV-2
W. P. #34 USN Battleships in Color
W. P. #35 Ticonderoga Class Cruisers

AIRCRAFT PICTORIAL SERIES

A. P. #1 Midway Air Wings
A. P. #2 SB2U Vindicator
A. P. #3 OS2U Kingfisher

Front Cover: USS NORMANDY (CG-60) off the French coast of Normandy, 1994, as part of the 50th. anniversary commemoration of the invasion of Hitler's Europe, on D-Day, 6 June 1944.

Back Cover: Atlantic Ocean, 11 June 2004, the guided missile cruiser USS VICKSBURG (CG 69) receives fuel from the fast combat support ship USS SEATTLE (AOE 3) in the Atlantic Ocean.